Albert W. Tipton

## The Electro Magnetic Principle of Creation

A new theory on the circulation of the blood, on the electro-magnetic

principle

Albert W. Tipton

**The Electro Magnetic Principle of Creation**
*A new theory on the circulation of the blood, on the electro-magnetic principle*

Printed in Europe, USA, Canada, Australia, Japan

Cover: Foto ©berggeist007 / pixelio.de

More available books at **www.hansebooks.com**

THE ELECTRO MAGNETIC PRINCIPLE OF CREATION.

# A NEW THEORY

ON THE

# Circulation of the Blood,

ON THE

## ELECTRO-MAGNETIC PRINCIPLE.

ALSO, OF

# THE TRANCE STATE.

———

BY

## A. W. TIPTON, M. D.

———

CHICAGO, ILLINOIS:
LIBBY & SHERWOOD PRINTING CO.
1892.

"Cuilibet in art sua crededum est."

"MULTUM IN PARVO."

Rochefoucaultd : — "Men of limited under-standings, in general find fault with everything which is beyond their comprehension."

Age by age, for centuries past, systems and practices that were considered scientific have dwindled away to nothingness, as time passed. To-day there are many things that are considered scientific in theories and practices, especially in the treatment of female diseases, that must meet with the same fate.

## INTRODUCTORY.

There has been a revolutionary action going on in my mind to which, twenty years ago, I said "Stop!" It stopped not, but the end is now. If a demonstrator of anatomy will give a natural, rational and practical solution of the reason why nerves uniformly accompany the arteries and not so uniformly the veins in the human system, which they have failed to do up to time of publication, I think I can prove, on a rational and natural basis, that the heart does not circulate the blood, but that the positive and negative electricities, generated from the air in the lungs and from them conveyed to the brain, thence distributed through the system to supply the natural demands of the body, circulate the blood and are the physical life and the motive

5

power of the body, acting in harmony with the spirit life; this may be likened unto the electricity, positive and negative, generated by steam, by water power or other known generative power of electricity, the principles of which are attraction and repulsion, push and pull, motive force or power. Everything is kept in motion by it as long as the power lasts or until the machinery wears out. So with the body. As long as you take air into the lungs, the machinery of the body will be kept in motion, until the machinery is worn out or destroyed by disease or outside influences, etc.

Spinal menengitis, congestions and inflammations of the human system can be explained and removed on scientific principles. Will wood and water run an engine without steam? Is not steam the motive power of the engine? Can you force water through a pump without a motive power? Will electric cars run without a motive power? Are not horses the motive power of our street cars? Can you run or keep in motion any kind of machinery without a motive power

of some kind? Is not the world kept in motion
by a motive power? Are not the lords of creation
kept in motion by a motive power? etc., etc.
And the man died for the want of air and the
heart failed for the want of motive power. The
engine failed for the want of steam and the
pump failed to deliver water for want of a motive
power, and every variety of machinery died for
the want of a motive power, and the world may
collapse from the same cause.

Is not the horse-shoe magnet in equilibrium,
electrically considered, and every molecule of
the magnet, a perfect magnet, electrically considered? And does not the same principle hold
good in all animate and inanimate matter? Is
not the electro-magnetic principle in stone the
life and cohesive principle of the inanimate substance? And would not the destruction of this
principle in nature destroy everything animate
and inanimate, in part or whole, in proportion to
the destruction of the substance or being? And
will not the horse-shoe magnet become deranged,
or the electro-magnetic equilibrium in a great

measure be destroyed, by dealing the magnet a heavy blow with a sledge hammer, or otherwise? And would not a person resort to electricity to restore the magnet to its original equilibrium, as the most efficient agent? And if not, why not?

When a person is in perfect health, are not the electrical conditions, positive and negative, in equilibrium or harmony in every organ or part of the body, like the magnet in equilibrium, electrically considered? And would not a heavy blow upon the person's head derange the polar and electrical equilibrium of all organs and parts of the body, through the nerves (motor and sentient), in proportion to the extent of the injury received by the blow, or otherwise? And would not electricity, properly and scientifically administered, restore harmony and an equilibrium in the system as in the magnet? And why not? Can animate matter or a person live without electricities, positive and negative, or north and south polarities? And if the motive power of electricity be great enough to raise a stone weighing thousands of pounds and convey

it from the earth to the top of a building 200
feet or more in height, run cars loaded with pas-
sengers, and do many other strange and wonder- ·
ful things, cannot the positive and negative electricities in equilibrium, generated from the air
in the lungs, circulate the blood? And is not
the physical life and motive power of the body
acting in harmony with spirit life? If not, why
not? Trance states can only be accounted for,
by me, on the principle that the physical life
departs or is divorced from the spirit while the
trance lasts. Cause, inco-ordination between the
electrical and spirit life, produced by the suspension of the physical forces. If not, why not?
(See page 73 for an explanation of the trance
state.)

# CONTENTS.

# PROGRESS OF ELECTRIC SCIENCE.

It is not a little interesting to note the progress by which electric science has advanced from its small beginnings, and some of the experiments employed for its development.

The first recorded discovery of the electric principle was by Thales, a distinguished Greek philosopher of Ionia, about 600 years before the birth of Christ. He observed it in amber, a resinous substance, which the most of you have probably seen in articles of adornment. Because he first found it in amber he called it *electricity*, from the Greek word ἤλεκτρον, electric, which means amber. With a curious and philosophical eye, Thales observed that amber, immediately after being briskly rubbed, as in *polishing*, drew to itself other lighter substances lying near it, such as feathers, bits of paper or papyrus, etc.

13

With him, however, this seems to have been regarded merely as a curious phenomenon. No practical results were deduced from it, nor do there appear to have been any further discoveries of consequence made of the presence or the properties of this remarkable agent, for the next two and one-half centuries, or thereabouts.

Then Thenphrastus, a celebrated Greek philosopher of Lesbos, who flourished over 300 years before Christ, detected it in the mineral called tourmaline. After him no advance worthy of note was made, in the knowledge of electricity, for nearly 2,000 years.

In the year 1600 of the Christian era, Dr. Gilbert, physician to King James I, of England, ascertained that a *large variety* of bodies, as opal, diamond, carbuncle, sapphire, quartz, amethyst, rock alum and several others, when excited by friction, attract to themselves other, light substances.

About the same time Ott Guericke, Burgomaster of Magdeburg, discovered electric repulsion, and himself, with two or three others—Dr. Wall, of England, among them—discovered also the electric spark.   Dr. Wall procured a long stick of amber, slightly tapering, and excited it by drawing it swiftly through a bit of woolen cloth in the hand, when frequent little cracklings were heard and each was accompanied by a small flash of *light*.   But by presenting his *finger* to the amber, thus excited, a much larger flash of light was elicited, attended by a loud and distinct snap.   Dr. Wall says, "It strikes the finger very sensibly, wherever applied, with a push or pull like wind.   This light and cracking," he sagaciously adds, "seem in some degree to represent thunder and lightning."   Bear in mind that this remark of Dr. Wall was made about 150 years before Dr. Franklin demonstrated the identity of electricity with lightning.

Again Stephen Gray, of England, in the early part of the last century, did considerable for the advancement of electrical science.   His principal experiments extended from about 1734.   He seems to have been a careful observer, especially of analogical relations.   Among his other discoveries, he was the first to note that knobs give off larger electric sparks than points, and was thus led to the following interesting though modest conjecture. He says: "There may be found out a way to collect a great quantity of the electric fire and consequently to increase the force of that power, which, by several of these experiments—*silicet magna componere parvis* (if it be permitted to compare great things with small)—seems to be of the same nature with that of thunder and lightning." This was about twenty-five years before Franklin's great discovery.

The *Leyden Jar* was the next important discovery in electrical science. This was about the middle of the eighteenth century, a few years only before Dr. Franklin used it so successfully both to give grandeur to electrical science and immediately to his own fame. The honor of this invention is held in dispute. It has been most commonly ascribed to Prof. Muschenbrock, of the University of Leyden, in Holland; but it has lately been claimed, with

much seeming confidence and apparently good
authority, for another native of Leyden, a Mr.
Cuncres, who, it is affirmed, first made the experiment by which electricity was accumulated
"on a glass vial called the Leyden Jar, because
the experiment was made by a native of Leyden,
but," the same writer adds, "the person who
made the discovery of, or observed first, the
phenomenon, was a Mr. Von Kleest, the Dean
of Commin; on the 4th of November, 1745, the
first shock was felt by this gentleman." (See
Dr. A. C. Garrett's Medical Uses of Electricity,
2d edition, page 67.)

Cuncres' experiment was made by placing a
metal vial or a thick brass wire in an apothe-
cary's glass vial and then charging the metal
with electricity.   Von Kleest says: "When this
is done very remarkable effects do follow."
Shortly after this Muschenbrock *repealed* the
experiment with a very thin glass bowl and got
a terrible shock.

These experiments led to the improved Leyden Jar, as we now have it; but its essential

principles were discovered first by the rude process that I have here stated. Muschenbrock says he received such a concussion in his arms, shoulders and heart, that he lost his breath, and it required two days to recover from the blow and terror; and he declared that for the Kingdom of France he would not take another such shock.

Winkler affirms that his body was thrown into such violent convulsions by it, and his blood into such an agitation, that he was obliged to take cooling medicines to keep off fever. At another time, as he says, the shock produced in himself hemorrhage at the nose, and the same effect was experienced by his wife, who was almost deprived of the power to walk.

The report of these remarkable effects excited the attention and curiosity of *all classes of people*. Nearly everybody wanted to experience the singular sensation, and a host of men, half taught, sought to fill their pockets by wandering over the country as electricians, to excite the amazement and gratify the curiosity of the people with the strange phenomenon. It is but justice to say here that the discovery of the Leyden Vial or Jar was due to the *previous* discovery by the prince of philosophers, Sir Isaac Newton, which was that electricity exerts its power, both of attraction and repulsion, through glass.

The electrical philosophers were entirely unable to account for the operation of the Leyden Jar until Dr. Franklin developed its true principles. His theory of it was at once adopted as satisfactory. Dr. Franklin, having in 1752 discovered the *plus and minus* of electrical states, or *positive* and *negative* polarities, observed that the outside of the jar was always negative, and this he *proved* by the following experiment:

He brought the free ends of the two conducting wires near to each other, these opposite ends being connected, the one with the outside and the other with the inside coatings of the jar, the jar being coated with metallic coverings, and between these free ends he suspended a small cork ball. The ball was immediately attracted and repelled alternately by each wire, swinging

like a pendulum between them, until the jar was
discharged.   This showed that there were op-
posing forces exercising control over the ball,
since it was always repelled from the wire which
it last touched, and, at the same instant, was
always attracted by the other.   This demonstrated the *fact* of the opposite electrical states ;
and these opposite states he showed to be *plus*
and *minus* or *positive* and *negative*, one side
losing what the other side gained or gaining
what the other side lost, by varying the experiment as follows :

He suspended a small linen thread from his
hand, near to a charged jar, and then observed
that the outside coating of the jar attracted the
thread to itself whenever he brought a finger of
the other hand near the wire that was connected
with the outside coating, the outside coating
plainly drawing in, by way of the thread, the
same quantity of the electric fluid that was taken
from the inside by way of the finger.   Here it
was proved, as Franklin considered, that the
outside and inside coatings of the jar were not

only in opposite electrical *states*, but also that in the changing of the jar the inside of it gains simply what the outside loses, and, consequently, that the difference between the opposite electrical states is only that of *plus* and *minus* or *positive* and *negative*.

Soon after the invention of the Leyden Jar, more than 100 years ago, a good deal of interest was excited in philosophic circles by the experiment of Dr. Watson, afterward Sir William Watson. This gentleman associated with himself several other Englishmen for the purpose of ascertaining, by trial, the distance to which the electrical action could be carried and the velocity of its motion.

The historian says: "On the 14th and 17th of July, 1747, they succeeded in conveying the shock across the Thames, at Westminster Bridge, by an iron wire, the water of the river forming a part of the chain of communication." It is said that one of the party held in one hand a wire, which communicated with the jar, and received the shock by dipping an iron rod, held in the other hand, into the water of the river. Of course, the wire that ran from the jar across the river, on the bridge, must have had its further end placed in the river, or in the moist ground, in order to complete the circuit.

On the 24th of the same month, Dr. Watson and his party, at two different places, one at New River and the other at Stroke Newington, succeeded in sending the charge over two miles. In one of these experiments, some 800 feet, or over a mile and a half of the distance, was through the water. As in neither of their experiments was any perceptible time occupied in the passage, they concluded it was instantaneous.

These experiments of Dr. Watson and his associates produced a profound impression on the scientific mind at that day, and did much to stimulate further researches in respect to the mysterious nature and laws of electricity. Yet how far they were from gaining the remotest conception of what is witnessed in our own time. Who could have thought of encircling a continent and spanning the ocean with, and making an intelligible messenger of that invisible agent that was found playing on the rubbed surface of a bit of glass or stick of amber?

In June of the same year (1752) in which Dr. Franklin discovered the distinction of positive and negative electrical states, and explained the principles of the Leyden Jar, he made in Philadelphia his celebrated kite experiment by which he " bottled up the lightning " and demonstrated the identity of electricity with lightning. He also proved the same fact again by an ingenious apparatus in his house, "connected an insulated iron rod with two bells, which indicated by their ringing that the rod was electrified." In this experiment he found the effects from natural and artificial electricity to be the same.

He also made the interesting discovery that the clouds are sometimes positively and sometimes negatively electrified, and that they often change their electrical states •during the same storm. At one time he found the atmosphere strongly electrified during a fall of snow.

Dr. Franklin was a man eminently practical, and delighted in subordinating philosophical research to purposes of utility. When, therefore, he found that he could, by means of a conductor, draw electricity from the clouds, it was, to such a mind as his, perfectly natural that he should seek at once to turn his discovery to a practical account. He did so, and soon electrical conductors were seen pointing upward from the tops of buildings and ships, designed to divert from those structures, to the earth or the sea, the spontaneous discharges of this fearful fluid of the heavens. His experiments in this direction proved a success and were the *first* application of electrical science to practical purposes of life.

Says an eloquent writer: " Hitherto electricity had not connected itself with any phenomena of wider range and importance than developed during the excitation of glass and other electrics." Astronomy has lifted the mind to the contemplation of the most august scenes in nature; magnetism connected her facts with the attraction of the great globe itself.

It has been conjectured, as we have seen, that the shock and spark of the electric machine were miniature effects of a more tremendous agent, but it was reserved for our own Franklin to raise electricity to its true dignity among the other branches of human knowledge. His discovery of the identity of the electric fluid with lightning was the step by which the change was effected.

The vulgar were astonished at the sight of fire brought down from heaven, and philosophers themselves were startled at the recollections that they had been amusing themselves with a thunderbolt and trifling with that terrible agent which had so often alarmed the intellectual and moral and convulsed the physical world. Indeed it has been forcibly said, in reference to that daring exploit of the American philosopher, " Human genius seems, on this occasion, to have made an impious excursion *beyond its mortal range*, and one victim was demanded to expatiate the audacious attempt; for, on the 6th day of August, 1753, Prof. Richmond of St. Petersburg was struck dead while making the experiment of drawing electricity from the clouds, as he bent forward to his instrument, to observe the degree or quantity of electricity which he had gained, he received a charge in his head which killed him instantly."

Since the time of Dr. Franklin many able men have taken up the study of electricity and prosecuted their researches with commendable zeal and much success until, with the greatly improved and diversified instruments that have been invented, this science, as related to both inanimate and animate nature, except in its *therapeutic* bearings, has attained a highly interesting and honorable position in the circle of sciences.

Irregular or fitful agitations of the needle were first observed in 1750 by Worgenthin, and in 1806 by Humboldt, who gave the accompanying phenomena the name of "Magnetic storms." There is still another variation of the magnetic pole of the earth for which no theory has fully accounted. The pole of the magnetic needle now points more than one thousand miles away from the geographical pole of the earth.  The needle pointed due north in 1660, in London, where the observation was made.  It then moved westward until 1818, when it was directed to a point 24° 27' from the pole; then it began to move back again and will point due north again in 1976, making a complete cycle in 320 years. Self-recording apparatus, now used in magnetic observatories, give daily and hourly reports of all magnetic variations, and when sufficient time has elapsed to secure enough observations from different parts of the world much light will undoubtedly be thrown on the cause of the earth's magnetism and its changes.

The earth, atmosphere and clouds form an earthern jar on an extensive scale, the earth and clouds representing the coatings of the jar and the air fulfilling the part of the glass through which the electricity passes by induction or discharge.  It is found that in fine weather the atmosphere is almost invariably charged positively; before rain it often assumes a negative state.  The rain that first falls is usually nega-

tive, although the atmosphere before and after
the fall may be positive.  Fogs, snow and hail,
if unattended with rain, are nearly always positively charged.  Clouds are always positive.

Electricity, like magnetism, has a maximum
and minimum intensity during the day that
may be traced to the influence of the sun, and
another during the night dependent on the moon.
There is also a yearly variation, dependent upon
the relative position of the earth and sun, atmospheric electricity having thirteen times as
great intensity when the earth is in that part of
its orbit nearest the sun, as when it has receded
to that part of its orbit most remote from the
sun.

There are also irregular or fitful disturbances
of the electric conditions of the atmosphere, accompanying the agitation of the needle during
magnetic storms.  These perturbations manifest
themselves, often simultaneously over land and
sea, over hundreds and thousands of miles, or
propagate themselves gradually, in a short space
of time, in every direction over the surface of

the land.   At these times occur brilliant displays of the Aurora Borealis, which are believed to be neither the cause nor the effect, but simply an accompanying phenomena of the electrical disturbances and due to the same cause.

To the German astronomer, Schwabe, is due the honor of recording daily observations during more than thirty years, by means of which he established the periodicity of these phenomena. He finds that they increase in number and frequency during a little more than five years, then decrease for the same period, occupying, to complete a cycle, about eleven years.   He also discovered that they coincide with the appearance of dark spots on the sun, although no one can say, from any evidence yet produced, that the storms are caused by the sun spots or that the sun spots are caused by the magnetic storms. Whether the sun is the source of electricity, or whether the magnetism of both earth and sun is derived from some common central reservoir of this force, still remains an unsolved problem.

The influence of terrestrial magnetism and

atmospheric electricity over health and disease is a subject of practical interest to every physician. That electricity is at all times present in the human body and that under certain circumstances it becomes manifest through sparks emitted from the person, as well as by other means, has long been known to all who have given any attention to the natural sciences.

Cecil relates an instance of a gentleman whose sensitiveness to atmospheric electrical conditions was so great that he was advised to insulate himself by wearing silk underclothing. So successful was this measure that he declared, "It made life another thing."

Dr. Hosford, of New Hampshire, reported in 1837 the following, which is interesting as describing a condition due to disturbed electrical conditions of the human body: "On January 5th of that year, during a brilliant display of Aurora Borealis (magnetic storm), a cheerful, intelligent lady, about thirty years of age, became suddenly and unconsciously charged with electricity, which she first discovered on attempting to pass her hand over her brother's face, when all the family were astonished to see a dis-play of sparks pass from her fingers to his face. This peculiar condition continued without dim-inution until the last of February, when it began to decline, and disappeared permanently in May. During its continuance, being greatly annoyed by disagreeable shocks on touching any con-ducting substance, such as kitchen utensils, needles or springs of chairs, every effort was made to relieve her, but neither the change of clothing from flannel and silk to cotton, nor any other device, gave her relief. She was not con-scious, from any internal sensations, of her peculiar power, though she could charge others, when insulated. She had never been confined to her bed by a day's illness, but had suffered for some months, during the previous year, with neuralgia, which permanently disappeared with the return of her normal electrical condition in May."

D. C. Woodman, of Paw Paw, Mich., (*Medical News*), reports the following curious case: "A patient, aged 27 years, can generate light through the medium of his breath, assisted by manipulations with his hands.  He will take anybody's handkerchief and hold it to his mouth, rub it vigorously with his hands while breathing on it, and immediately it bursts into flames and burns until consumed.  He will strip, and rinse out his mouth thoroughly, wash his hands, and submit to the most rigid examination, to preclude the possibility of any humbug, and then by his breath blow upon any paper or cloth, and envelope it in flames.  He will, when out gunning and without matches and desirous of a fire, lie down after collecting dry leaves and by breathing on them start the fire, and then coolly take off his wet stockings and dry them.  It is impossible to persuade him to do this more than twice a day, and the effort is attended with the most extreme exhaustion.  He will sink into a chair after doing it, and, on one occasion, after he had set a newspaper on fire, as narrated, I placed my hand on his head and discovered his scalp to be violently twitching, as if under intense excitement. He can do this at any time, no matter where he is, and under any circumstances." Dr. Woodman has repeatedly known of his sitting back from the dinner table, taking a swallow of water, and by blowing on his napkin, at once setting it on fire. He says that he first discovered his strange power by inhaling and exhaling on a perfumed handkerchief that suddenly burned while in his hands.

The following case was reported by Dr. C. A. Leal (Journal of Nervous and Mental Diseases, October, 1876): "A strong man was able to light the gas with ease after a few frictions with his finger. He was recommended to rub his wife, who was suffering with neurasthenia. She recovered, but he became morose and restless, and complained of a disagreeable feeling when his wife touched him. He finally recovered but was not able to engage in his former business."

There are occasionally reports, through the newspapers, of the effects of magnetic storms upon invalids and nervous people—faintings, spasms, palpitations, etc., having occurred when the subjects were not aware that there were any electrical disturbances. Many invalids are enabled to foretell changes in the weather by the aggravation or amelioration of their disease, and their change of symptoms will be found to correspond with the change in atmospheric electricity from positive to negative or vice versa, which immediately precedes or follows storms. The daily and nightly rise and fall of the grave symptoms attendant upon many acute diseases correspond very nearly with the variations in terrestrial electricity, and are recognized as occurring with such regularity that the experienced practitioner can often readily predict the condition of the patient for hours in advance.

Certain groups of diseases are influenced by the seasons. The greater prevalence of lung diseases during the winter months and of bowel complaints during the summer months cannot be fully explained by the differences in temperature and diet; while of the epidemics, small pox is recognized as a winter disease and cholera as a summer disease.

Although no feasible theory for this has hitherto been advanced, so far as known to the writer
of the paragraph regarding the agitation of the
needle and those subsequent to it (W. S. Haines,
M. D.), the variation in atmospheric pressure
undoubtedly exerts considerable influence over
the state of health, but this very variation of
pressure would of itself greatly influence the
electrical conditions of the human body.

From the days when the Greek slave of
Anthero was subjected to the shocks of the tor-
pedo or electric fish to cure him of his infirmity,
and the Grecian women and children wore beads
of amber in the belief that its mysterious soul
would exert a healing influence over their diseases, mankind has, from time to time, sought in
electricity a panacea for all human ills.  At times
its champions have made the most apparently
impossible promises for it, which, failing to be
fulfilled, have thrown disfavor upon its power,
principally on account of the ignorance of those
who used it, and it would pass into obscurity
forgotten by the public until an opportunity pre-

sented to again call attention to it as possessing almost miraculous properties.

During the period it has been undergoing these alterations in popular favor, a few ardent lovers of science have quietly pursued its investigation with such a wealth of reward in knowledge of its properties and its possibilities that it has been recently remarked by many writers that probably " the age of *discovery*, so far as regards electricity, is past, and we have actively engaged upon the age of the *practical application of principles* recently demonstrated." To such an extent has this been done as to enable a skilled electrical therapeutist to know what sort of effect will be produced on acute and chronic diseases to a certainty before making an application of electricity, just as it is known what effect will be produced by a knife when applied to the flesh. Electricity is a mechanical agent like the knife, and its therapeutic value can be learned by extensive experience with the agent on the principles taught in ELECTRIC MEDICATION.

# VITAL FORCES, ANIMAL AND VEGETABLE.

Upon these points I must be permitted to offer a few words. Of the animal kingdom I regard the nervous fluid or "influence," popularly so called, as being the principal of *animal vitalization*, the life force and a modification of the ELECTRIC FORCE. It is, I think, pretty generally conceded at this day that the "nervous influence" is electrical or electro-magnetic. There are some alleged facts, and other certain facts, which go far to sustain this view.

It is said that if we transfix with a steel needle a large nerve of a living animal, as the *great ischiatic*, and let it remain in that condition a suitable time, the needle becomes permanently magnetized. So, too, if the point of a lancet be held between the severed ends of a newly-divided large nerve for some length of time, that point, as I have heard it affirmed on what appeared to be good authority, becomes magnetized; although I have not attempted to verify either of these cases by experiment. However, admitting them to be true, the metal is charged with the nervous fluid.

But the fact on which I, myself, chiefly rely for evidence of this identification, being almost daily conversant with it in my practice, is this: The "nervous influence" obeys the laws of electrical polarization, *attraction and repulsion*. When I treat a paralyzed part, in which, to all appearances, the action of the nerve force is suspended, I have but to assume that this force is electric and apply the poles of my instrument accordingly, and I bring it in from the more healthy parts along with the inorganic current from my machine. Forcing conduction through the nerves, by means of my artificial apparatus, I rouse the susceptibility of the nerves until they will normally conduct the "nervous influence" or electro-vital fluid, as I term it, and paralysis is removed.

Again, if I treat a part in which the capillaries are engorged with arterial blood, I have but to assume that the affected part is overcharged with the electro-vital fluid, through the nerves and the arterial blood, and I so apply my electrodes, according to the well known electrical law, as to produce mutual repulsion, and the inflammatory action and congestion are sure to be repressed. I manifestly change the polarization of the parts. This thing is so constant and regular that I am entirely assured, before touching the patient, what sort of effect will be produced by this or that arrangement in the application of the poles of the instrument. If I desire to increase or diminish the nervous force in any case I find myself able, on this principle, to produce the one effect or the other at will. Hence, I say, the nervous influence obeys the electric laws, just as does the inorganic electricity.

I find this subtle agent not in the nerves only, but in muscle and blood—more especially in arterial blood. Indeed it seems to pervade, more or less, the entire solids and fluids of the animal system, and wherever it exists its action is just that of electro-vital force. While, therefore, I cannot affirm the identity of animal electricity and vitality, the theory of their identification, to my view, best accords with manifestations under correct therapeutic treatment, and I am unaware of any established fact disproving it.

*Vegetable vitality*, also, I regard as another modification of the electric force. The fact has been proved, by repeated experiments, that galvanic or electro-magnetic currents, passed among the roots of vegetables, cause a quickening in development of plants to a degree that would be deemed incredible by almost any one who had neither seen nor learned its rationale. I have seen it stated, on authority which commands my credence, that by this process lettuce leaves may be grown, within a few hours, "from the size of a mouse's ear to dimensions large enough for convenient use on the dinner table."

# EXTENT OF ELECTRIC AGENCY.

When we have settled upon the position, that electricity of the heavens and that of the artificial machine are identical, and that their identity is essentially one with electricity, galvanism, magnetism, the electro-vital fluid of the animal kingdom and the life force of the vegetable kingdom, it requires no extravagant imagination nor remarkable degree of enthusiastic credulity to suppose that all the forms of physical attraction and repulsion are due, under God, to the diversified modifications of the all-pervading agent, ELECTRICITY.

Indeed, for myself, I feel no hesitation in expressing it as my belief that electricity, in one phase or another, and controlled only by *will*, is the grand motive power of the universe. I believe that, in the form of electro-vital fluid, the Creator employs it as his immediate agent to carry on all the functions of animal life; and that, in respect to voluntary functions, He subordinates it. as a servant, to the will of the creature, to effect such cerebral action and such muscular contractions as are demanded by the creature's volition.    I am disposed to think that, by the omnipotent power of His will, He controls and uses electricity, in its various modifications, as the immediate moving force by which He accomplishes all the changes in the physical universe.

It is fast becoming a generally received opinion among modern savants that every body in in nature is electro-magnetic, more or less, and that all visible changes are but the result of changing poles.   Chemical affinities and revulsions are believed to be only the more delicate forms of the electrical attraction and repulsion, the ultimate particles of matter, no less than matter in masses, being subject to the control of electrical laws.

The imponderable agents, light and caloric, under the ingenious tests of scientific scrutiny,

are beginning to give some very decided indica-
tions of being simply electric phenomena.  Indeed, the doctrine or theory that supposes caloric
to be simply atomic motion is even now very
generally accepted by the scientific world.   And
that motion in the atoms of a body which causes
in us the sensation of heat, is probably electric
motion   And permit me to observe that though
the operations of nature seem, at first thought,
to be wonderfully complex and mysterious, yet,
if the views here presented be correct, the mar-
vel is changed, and we are brought to a profound
admiration of the simplicity of the means by
which the Almighty conducts His material operations.   A single agent made to perform pro-
cesses so infinitely numerous, diversified and
apparently complex!  How amazing!  Majestic!
Like the mind of God!

# WHAT IS MAN?

My answer to the above is as follows : Man is a three-fold being, composed of a body material, a body electrical, and a spirit rational and indestructible.

I. THE MATERIAL BODY. This is composed of various metals, earths, carbon, phosphorus and gases. I need not go into a representation of their multiplied and curious combinations to form the many parts of the body complete; but there are the ultimate elements, and a most superb and wonderful structure they here compose. Yet, notwithstanding all the manifest skillfulness of its contrivance and the power of its accomplishments and the beauty of its execution, it were a useless display if unaccompanied by the invisible agents which compose the other two grand constituents of man, to-wit: it would quickly fall into decay, as we see it when deprived of them, and would be resolved into its original elements again. But to our gross material bodies the creator has added:

II.  THE BODY ELECTRICAL.  By this I mean that which commonly has been termed "nervous influence," " nervous fluid," " nervo-vital fluid " and "nervo-electric fluid."  I object, however, to each and all of these designations. They are too restricted and specific.  They all seem to imply that it is an influence which appertains especially to the *nervous* system; whereas the entire system is under its pervading force.  I do not doubt that its chief action is in and through the nervous system, but it also pervades, and, as I think, vitalizes the whole body.  The nervous system seems to be created as one principal means for its replenishment and to serve as the medium of its ministrations to the body at large. I choose to term it *electro-vital fluid* or *electro-vitality*.  My reasons for so designating it are the following :  (1.) It is demonstrably electrical in its nature.  (2.) It appears to be identified, or at least immediately connected, with the vitalization of the body. (3.) I wish, by its name, to distinguish it from *mental* vitality, or the vitality of the spirit.

Whether, as a peculiar manifestation of the electrical principle, it vitalizes by its own nature and action solely, or whether it be charged with another mysterious element, a life force, and vitalizes by ministering the latter to the material organism, I will not positively affirm. Whichever it be, the name I assign to it seems sufficiently appropriate. But I strongly incline to the theory that this electro-vital principle, by virtue of its own nature, vitalize the system. In other words, I am led to think that God makes it the immediate agent of vitalization, having constituted it the *vis vitae* of both the animal and vegetable kingdoms. Nor does this idea, as I conceive, necessarily conflict at all with the doctrines of cell-life, as maintained by the best physiologists of the present day. I also sometimes style this electro-vital element the *body-electrical*, because it is certainly an entity, co-extensive with, and, in greater or less force, wholly pervading the visible, material body.

At this point I will take the liberty to intro-duce, though somewhat digressively, a few thoughts on the *distinctions of vitality or life.* There are, as I suppose, the following kinds of life: (1.) Spirit life. (2.) Moral life. (3.) Electric life.

(1.) There is spirit life. And here are to be made several divisions. (*A.*) Uncreated spirit life. This is the life of God. Of the nature of the Divine Essence we know nothing; yet that God is a real, living entity we do know. My own conviction is that the Divine Essence and the Divine Life are identical; that God, a spirit, is necessarily infinite, conscious *vitality*, the volun-tary originator of all existences beside Himself. But as to what is the essential nature of this vitality, this eternal spirit life, we can have no conception, only that this life is God. (*B.*) Created spirit life. Here we make another sub-division: (*a.*) The life of created mortal spirits, which is a rational, intelligent entity, representing the spirit of man and of unembodied, created intelligences above him, created as it pleased God "in his own image" a living, indestructible essence; and, as I suppose, its essence and life are the same. (*b.*) The life of created mortal spirit, as the spirit of the beast, of its intrinsic essence we are also necessarily ignorant, yet of its attributes we know that it has consciousness, sensibility and will; of its life we know as little as of its essence, both of which, however, as I conjecture, are also one and the same, the spirit substance being itself essentially vital.

(2.) We pass next to Moral Life. This life is identical with holiness, the very opposite of that defilement that characterizes moral death, which is a state of sin. But let me again subdivide. (*a.*) As the to moral life of God—it consists in His infinite moral purity, His veracity, justice and benevolence, or love qualities, which in their combination make up His holiness. (*b.*) The moral life of man, as also of other rational creatures, consists in his sympathy of spirit with God in respect to those pure qualities which constitute the Divine holiness.

(3.) Finally there is Electric or Physical Life, but here again there are varieties. (*a.*) There is animal life as of man and the lower animals, which I have already represented as constituting the electro-vital force. (*b.*) Vegetable life, which is another modification of the same essential electro-vitality.

But now to return to the physical or animal life of man, the electro-vital element. While this is in such immediate relation to the visible, on the one hand, it holds also, on the other hand, an immediate relation to the mental part, both of man and of the other animated beings of earth. It serves to transmit, through the nervous system, to the mind, all sensations and impressions from the outer world. It, moreover, receives from the mind the action of its volitions and imaginary conceptions, and conveys, through the nerves, the impressions or impulsions thus obtained to the various parts of the body, and there secures the fulfillment of the mind's behests. It appears to be only in this way that communication is had between the mind and its

outer body. The natures of the spirit and of
gross matter are so unlike that it seems im-
practicable for the mind and body to come into
immediate mutual relations or to act reciprocally
without the aid of a medium, ethereal, semi-
material and semi-spiritual, such as is the electro-vital fluid. The Creator has accordingly
provided this mysterious, invisible medium between the two, and thus, in a degree, extended
man's likeness to Himself by making him a
trinity in unity.

III. THE MIND OR SPIRIT. This is immeasura-
bly the highest and most important constituent
of man. His body material will fall back to
dust; his body electrical may be reabsorbed in
the great ocean of natural electricity that fills
the earth and the heavens; but his spirit is im-
mortal. His spirit, made in the Divine image,
lives and acts, thinks and feels, independently
of every other existence save Him from whom
it came. While in connection with its visible
body, its good or ills, its woes or bliss, has indeed
much to do with its bodily state; but, when

separated from this body, its high and more in-
dependent existence is at once asserted, and then
its good or ill are determined by its Author only
in accordance with the workings and affections
within itself. A spiritual and indestructible
being, like its Creator, it can never cease to be
while He exists.

But our present concern is with the mind in
its relation to that electro-vital medium between
it and the body, and to the body itself. The
mind's influence upon both of these parts of the
entire man is truly wonderful, although perceptible mostly on the material body. Few persons
are aware how much the state of the mind affects
the bodily health, although the degree is often
very great. Yet this is done by the mind's
action, first on the electro-vital functions, and
through these, by way of the nerves, upon the
bodily tissues. Changes in the mental state
will frequently produce change of polarization
in the physical organs, and thus determine in-
falliby the matter of health or disease. So, too,
the condition of the bodily health will often

determine irresistibly the mental state. What-
ever bodily changes affect the polarization of the
electro-vital medium, in any part of the organism,
do thereby produce corresponding change in the
mind.

These views of the reciprocal action between
mind and body, through the medium of the
electro-vital element, may serve to explain those
psychological wonders exhibited in the cure of
diseases by the imagination, as well as in diseases
and even death induced by the imagination. I
would like to unfold and illustrate this bearing
on the subject, and also, in the light of it, to show
the philosophy of one's mind acting intelligibly
on another mind with, and even without, the aid
of the physical organs, as is sometimes seen in
the feats of mesmerism.

There is another thought which I will offer in
this connection. I maintain that all functional
action of our bodily organism, *ab initio*, is conducted by thinking mind, through the medium
of organic electricity or the electro-vital fluid.
Every organ as a whole and every life-cell in

detail, is charged with this active principle. I
believe that every one of them is controlled and
guided incessantly in its propagatory, organizing
and entire functional force by intelligent mind,
acting through this wonder-working agent, the
electro-vital fluid. In respect to our voluntary
exercises, this organic electrical force is made
subject to our own mental activities, and exe-
cutes its office upon the bodily organism mainly
through the medium of the nerves. But, as regards all the involuntary functions, I believe
that control is exercised directly by the omnis-
cient and all-pervading God, directly in accordance with His own established laws.

Once more let me remark of the mind that
*consciousness*, *sensation* and *will belong to it alone.*
The *body* never feels nor thinks, nor does the
organic electricity within it. The popular idea,
especially with the uneducated masses, is that,
if a man burn his finger, it is the finger that
smarts. But this cannot be true. Pain cannot
exist except where consciousness is, and there is
no consciousness in the finger nor in any ma-

terial part. Only the mind is conscious of existence even, and hence only the mind can be conscious of pain. If a limb be paralyzed by interrupting in any way the flow of the electro-vital fluid through its nerves and thus depriving the mind of its medium of communication with it, you may burn that limb to a crisp and the subject will feel no pain. When you burn your finger or break your arm, you disturb the action of the electro-vitality in the injured part, deranging its poles. This electric agent instantaneously communicates its disturbance along the nerves to the brain, where it reports to the mind and locates the disturbance. The conscious mind takes cognizance of the fact and feels distress.

# THE LOWER ANIMALS.

It may, by some, be objected that, if we regard sensation as existing only in the mind, as affirmed above, we must concede mind to the lower animal tribes, since they are subjects of consciousness, sensation and will, as ourselves. I admit this necessity and unhesitatingly take the position, as has already been done in classification of minds, that the lower animals are, in fact, endowed with a something, higher and more spiritual than their material bodies or their animal vitality, something which bears distinguishing characteristics of *mind*. I would not, however, be understood to imply that they possess *all* the characteristics of our minds, even in a rudimentary degree, for I do not believe they do. My theory does not accord to them either reason or immortality. Yet, in respect to the latter, my views are less discursive and my utterances usually more reserved. But I think their minds may and probably do perish with their bodies. Nevertheless, the existence of consciousness, sensation and will, in any orders, does evidently presuppose some sort of mental constitution. And such mental structure, in them as well as in us, must be distinct from and superior to the animal vitality, compelling service from the latter and using it as a medium for communicating with the body and with the outer world in general.

# THE VEGETABLE KINGDOM.

As to the vegetable kingdom, there is here, so far as we can discover, only a quality of principle, viz: the material body and a modified phase of electro-vitality. These component parts appear to sustain each other in the vegetable relations, quite analogous to those of the corresponding parts in the animal. But here the mental part is wanting, and consequently there is no consciousness, sensation or will, and the electro-vital action is guided, in its elaborate and beautiful operations, for the forming and development of the plant, and in all its vital functions, by the all-pervading mind of God.

# NATURAL POLARIZATION OF MAN'S PHYSICAL ORGANISM.

The electro-vital fluid, in the animal economy, is subject to the same principles of polarization as the magnetic current from the artificial machine or the magnetism of the bar magnet. In the material organism of man, the great nerve centers, the brain, the spinal cord, and the ganglions, appear to act the part of fixed magnets, charged with the electro-vital fluid. Indeed, there is much reason to believe that this fluid is elaborated within the nerve centers, more especially within the brain, from the inorganic electricity of the outer world, which is supplied through the lungs in respiration and conducted thence to these laboratories, by a remarkably interesting process. These nerve centers, viewed as magnets of electro-vitality, require to be regarded as having each a positive nucleus in every direction to the surface of the medullary organ, radiating, as it were, from center to periphery. And the nerve lines and ramifications which issue from these great nerve centers are polarized evidently in the same way, the electrovital fluid being disposed with its negative ends to the positive surfaces of the nerve centers and its positive ends to the "vital organs" and especially to the surfaces of the organism as a whole. There are many other polarizations in the human system, subordinate to those mentioned above. The foregoing is preparatory to the introduction of the theory of the circulation of the blood.

# THEORY OF THE CIRCULATION OF THE BLOOD.

## A NEW THEORY ON THE CIRCULATION OF THE BLOOD, ON THE ELECTRO-MAGNETIC PRINCIPLE.

The air is taken into the lungs and with great rapidity is combined with the elements of the venous blood, thereby liberating a great quantity of electricity, which now, as free electricity, takes to the oxidized arterial blood, as its conductor; just as, in our electrical machines, when, by the combination of the acid with the zinc, electricity is liberated, it rushes to the conducting platina plate. The presence of this free electricity in the arterial blood is believed to be indispensable to complete its essential character as arterial blood. The lungs are the great battery of the system; in them, chiefly, the system gets its magnetism, obtained from the elements of the venous blood, especially from metals and carbon, by action of the oxygen of the air upon them, in the function of breathing. The brain nerves are the agents or instruments of the distribution of the blood to the system in general.

Allow me to ask you to bear in mind the cardinal principle of electrical polarity, that positives repel positives, while positives and negatives attract each other. You inhale your breath, as a vital function, by the action of the electrical principle sent out from your brain to the organs employed. That breath is full of oxygen. The venous blood, in the lungs, is composed largely of oxidizable elements, as the carbonaceous and the metallic. The inhaled oxygen, coming in contact with these elements in the lungs, and having a powerful electrical affinity for them, combines chemically with them. This combination or oxidization of the sanguineous fluid changes it from the dark venous to the red arterial blood, and, by charging it *somatically* with the magnetism which was liberated in the process of oxidization, makes it strongly positive. But two positives repel each other. The lungs, being constantly replenished and filled with this electricity, set free by the action of the oxygen on the venous blood, are also positive. The lungs, therefore, and the arterial blood, repel each other; their mutual repulsion necessitates and compels their separation; the one or the other must give place; the lungs cannot retreat, and the blood must. Its only course to get away is through the pulmonary veins, and it rushes through them into the left auricle of the heart. But it is yet within the sphere of the mutual repulsion between it and the lungs, and, being also crowded along by the succeeding current from the lungs, it is urged onward.

The auricle, under the influence of the electro-vital force, performs its proper function, contracting on the blood within it. The blood shoots through the passage down into the left ventricle, where it still feels the powerful repulsion of the lungs and the propulsion of the advancing current behind, and here also the

ventricle, actuated by the electro-vital force, performs its duty and contracts with much force
upon the blood. This action of the heart, sup-
plemented and sustained by the forces in the
rear, drives it to the extremities of the arterial
system. Here the propulsion from behind is
aided by the electrical attraction of the capillaries
in the advance, of which Prof. Draper speaks.
Hence the blood is both crowded and drawn forward into these minute vessels.

But all along its way, so far, there has been a
friction between it and the arterial walls, causing
it to part with more or less of its electricity. In
the capillaries this process is completed. The
chemical changes there, which, by the way, I
think to be really electrical changes, take from
it the balance of its oxygen and with that its
relatively positive electricity, and so change it
into negative venous blood. Now its electrical
relations to the lungs are changed. Here in
these systematic capillaries, it reaches the central
tral point of its circuit. Being now electrically
negative, it both attracts and is attracted by the

positive lungs.  It therefore rushes to them by the only route that is open to it; that is, through the veins, heart and pulmonary artery.

In the lungs it enters the pulmonic capillaries, where it again comes into contact with the oxygen inhaled as an element of the air.  By the above mentioned action of the oxygen, it is immediately charged with electricity and thus becomes positive again, when the old repulsion between it and the lungs is renewed and it is again repelled to the left auricle of the heart, driven through it into the left ventricle, whence it is thrust out into the aorta by the heart's mechanical contraction and urged on by magnetic repulsion through the arterial system as before.

Thus it continues its rounds ; on the one side being driven off to distribute through the system its nutritive and life-sustaining freight, which had been received along the veins from the absorbent vessels of the alimentary canal and in the lungs ; and, having delivered its charge, being called back, on the other side, by a resistless power, to the lungs to be freighted anew. Such is the circulation of the blood; and such are the forces, as I humbly conceive, by which, in its circulation, the blood is impelled.

At this point, the question may very properly be raised: What becomes of the electricity or magnetism that is given off from the arterial blood in its passage from the heart through the aorta and the arterial ramifications? A very important inquiry and one the answer of which should be well understood. This question brings under consideration one of the many beautiful exhibitions of contrivance, in the structure of the human body, which proclaim with clearness and force the existence of a conscious, intelligent God, the Contriver and Creator of all. I refer to the well known fact that the arteries are invariably accompanied, in all their flexions and ramifications, by, at least single, and much of the way double, lines of closely attendant and interwining nerves; a fact which is not true of the veins.

Now for what purpose, do you suppose, is this special nervous attendance granted to the arteries and no the veins? That at least a single line of nerves should be served out to the arteries would seem to be for the purpose of exciting, by means of the electro-vital fluid, through the nerve filaments that penetrate the arterial issues, those contractions which co-operate with the contractions of the heart to facilitate the circulation of the blood. But for what purpose is this duplicate nerve line that accompanies the vessels containing arterial blood? Most manifestly, as it seems to me, for the purpose of taking up, by induction through its filaments, which entwine about those blood vessels, that electricity which is discharged from the arterial blood by its friction against the walls of the arteries, and conducting it to the great sympathetic, and thence, by anastomosis, to the spinal cord and up to the brain; there to undergo some secret elaboration in conjunction with the mind or spirit, converting it into electro-vital fluid, to be sent forth again as the vital force of the whole organism.

You will remember that the arterial blood is highly charged with electricity, and therefore strongly positive. But when it has passed from the arteries to the veins, it has lost its magnetism and becomes negative, venous blood. This loss is, and must be, largely sustained along its passage, since there is considerable friction there. It is true that the inner surface of the arteries is highly polished, but the same is true also throughout the whole circulatory system, and yet, in the capillaries and other small blood vessels, as seen in the web of a frog's foot, there is so much friction that the red corpuscles are visibly retarded in their motion along the *parietes*, while those in the center move by them and pass along with greater ease and speed. So in a river the friction of the water with the banks necessitates a slower current along the edges than in the middle of the stream.

Precisely this is seen in the blood vessels. No matter, therefore, how polished are the inner walls of the arteries, there is a friction there with the flowing blood, and that friction must discharge the electricity of the blood throughout the arterial course. This electricity, I doubt not, is disposed of in the manner that I have already stated. Thus I have endeavored to indicate to you that most admirable replenishing apparatus by which is kept up, in health, a continual and abundant supply of the electro-vital fluid to the system, while every exercise of mind and of body is put forth at an expenditure of this vital element, to exhaustion, prostration and death.

# THE TRANCE STATE.

The trance state is a condition produced by the suspension of physical life—cause, inco-ordination between the electrical life and mind, or spirit life—in which condition the person is supposed to be dead, yet lives, and in many instances is buried alive, as numerous undertakers and others will testify. The state or condition simulates Catalepsy more nearly than any other known disease, and the same cause of the disease may be attributable to the same or similar causes as those of Catalepsy. This trance state or cataleptic condition may continue for weeks, without a sign of life in the body, when, from some unknown cause the physical life reunites with the mind or spirit life, and the person becomes a living, moving being. From a Bible standpoint, God made man from the dust of the earth and breathed into his nostrils the breath of life and he became a living soul or spirit.  Now, when in the trance state, the physical life is suspended or departs from the body, the spirit life remains in the body, which in itself is unchanged and will remain so as long as the trance lasts.  This proves the authenticity of the Bible statement, that man is endowed with a living spirit, beyond a question of doubt.

When the spirit departs from the body it returns to God, who gave it, and the body returns to dust, whence it came.  This may be likened unto the separation or removal of the electricity from our street cars.  First, the electricity, or physical life was suspended or left the body; eventually the mind or spirit left the body, which ended all until the resurrection of the dead ; just as the electricity leaves the car standing on the track until the motorneer turns his electricity on to put his car in motion.  Now, if any person has a more practical, natural and rational theory on the subject, I should be pleased to hear from him.

# ANALYSIS OF THE PRACTICE OF MEDICINE.

## GIVING THE REASONS WHY ELECTRICITY SHOULD BE USED IN CONJUNCTION WITH MEDICINES, ACCORDING TO THE PRINCIPLES TAUGHT IN MY WORK ON ELECTRICAL MEDICATION.

The practice of medicine is based on the theory that medicines act, on diseased organs, through the medium of the nervous system, circulation, etc. The experiments with medicines have been made on persons who were in a normal or healthy condition, and if administered when in this condition, they will have the effect they are accredited with. If the mediums are one-fourth less healthy, on the same principle, the medicines will have one-fourth less effect. If they are one-half less healthy they will have one-half less effect.   When used without the aid of electricity, they will rarely, if ever, effect a cure. Less medium, less action.   No medium, no action.

# INFLAMMATION AND WHAT IT IS.

The term inflammation is derived from the Latin words "*in*" and "*flamma*," in flame, fire, combustion, because of the burning pain attending it and the appearance of the parts affected. This is correct as far as appearances go. But what it is, and how to get rid of it, are the concern of the physician, if he has a severe case of inflammation or congestion on his hands. Inflammation is an excess of nerve force, or electricity and blood, in the parts inflamed. Electricity is the fire, while blood is the fuel; hence, also, the philosophy of redness, pain, unnatural heat, etc.

Now for the evidence of the correctness of the theory. In severe cases of congestion or inflammation, you will find a coldness of the surface and extremities in every direction from the local congestion or inflammation, for the reason that the electricity and blood have receded from the surface and extremities to the parts inflamed, impairing the mediums through which medicines act. It is a law of nature that a greater force will repel a lesser one. Therefore electricity, scientifically administered, will force or repel the excessive amount of electricity from the inflamed parts to the surface and extremities, from which it had receded, re-establish an equilibrium in the system and maintain it; thereby preventing all possibility of death from inflammation, if applied before mortification ensues. This it has not failed to do, in my hands, for the last twenty-three years, excepting in two cases. A partial record of my practice for twenty-two years will explain. Price 10 cents.

# LAURA BRIDGMAN'S BRAIN.

The brain of Laura Bridgman, the famous woman who lived her alloted years, devoid of sight, hearing, speech, smell and taste, brought some time ago to Clark University for examination, has just told its story. The result of the investigation proves that the peculiarities were due solely to arrested development in the portion relating to the disused senses. Up to the time of the girl's illness, when she was two years old, the brain developed normally. After that it grew unevenly. It is a well known and undisputed fact that the use or disuse of certain portions of the body, or of certain sets of muscles, results in a marked development or lack of development of these portions or muscles. The theory quite naturally was advanced that the development of the brain tissue depended, to a considerable extent at least, on the use or disuse

of senses directly connected with and relating
to the cells of the brain; There was never so
good an opportunity to test its truth as in the
case of Laura Bridgman, a healthy woman, with
an originally normal brain, but which lacked
development, living to be nearly 60 years old,
exercising to a very considerable extent the
powers left her, and never moody or despondent.

The weight was but slightly less than that of
the entirely normal brain.    Both hemispheres
were developed alike.    The extent of the cortex
(which receives and imparts sensations) was, if
in any way unusual, somewhat less than the
average brain.    All of the affected cranial or
brain nerves were small, and the regions of the
cortex associated with the defective senses and
motor or articulate tongue-speech were partly or
peculiarly developed.    In general the entire cortex was thinner than in the normal brain with
which it was compared.    The nerve terminations
of the nose and eyes were destroyed or highly
disordered and there was great destruction of
the middle ear and of the nerves connected with

the organs of taste.   The most striking and conclusive feature, however, was the condition of the parts connected with the nerves of sight. The right eye of Miss Bridgman remained useful, to a slight extent, some time longer than the left.   This resulted in developing that portion of the brain connected with the right eye to a greater extent than that connected with the left.   This is sufficient proof in itself that the development depends upon the use of the organ.—*Clipping from newspaper.*

### EXPLANATION.

Analysis of the case goes to prove that during her sickness, the cortex (which receives and imparts sensations) was engorged with electricity and blood; this, owing to the suspension of the electro-magnetic forces from remote parts to the cortex of the brain, was the cause of inflammation, which continued long enough to disarrange or destroy the conductors or nerves leading from the cortex to the diseased parts named, leaving a deathlike state, upon subsidence of the inflammation, in the conductors from their origin in the brain to their termini. Had electricity been scientifically administered at the time the cortex was engorged, and the inflammation dispersed, there would not have been a diseased condition left in her brain. This I positively believe from actual practice or experience in treating congestions and inflammations of the brain. (See Electrical Medication, page 22, 1888, for treatment of the brain in congestions and inflammations of the same.)

# LIFE AND DEATH, ELECTRICITY AND WATER.

THE HUMAN SYSTEM, IN HEALTH AND DISEASE, COMPARED TO A POND OF WATER WITH AND WITHOUT A LIVING STREAM OF WATER RUNNING THROUGH THE SAME.

Suspension of nerve force in the system suspends the circulation of the blood and weakens the muscular fibers and capillary system in proportion to the suspension of said nerve force. The whole human structure will become deranged and diseased to the same extent of the suspension of nerve force in every organ and part of the body. In proportion to the amount of pure water running through the pond will the stagnant and impure water be purified.

Stagnation in the system, as in water, is death-like. Electricity, scientifically administered, will purify and restore action in the system on the same principle that a stream of water running through a pond of stagnant water will purify the same. See Electrical Medication for general principles.

CONTENTS: Polarization, The Electric Circuit, Polarization of the Circuit, The Current, Modifications of Electricity, The Vital Forces—Animal and Vegetable, Extent of Electric Agency, Theory of Man, The Lower Animals, The Vegetable Kingdom, Natural Polarization of Man's Physical Organism, Electrical Classification of Diseases, Philosophy of Disease and Cure.

---

PRINCIPLES OF PRACTICE.

Polar Antagonisms, Importance of Noting the Central Point, Distinctive Use of Each Pole, The Use of the Long Cord, The Inward and the Outward Current, Mechanical Effect of Each Pole, Relaxed and Atrophied Conditions, General Directions of the Current, Treating with Electrolytic Currents, Positive and Negative Manifestations, Healing Diagnosis.

teaches how to treat numerous acute and chronic .
diseases with electricity. It teaches how to control inflammation to a certainty by using electricity in conjunction with medicines, thereby saving the lives of many that would perish or die without it. Over thirty-eight years' experience with medicines, twenty-three years of the time a clinical experience with electricity, justifies the above assertions of the author.

86

# REFERENCE REVIEWS.

J. J. Lawrence, A. M., M. D., editor of *Medical Brief*, St. Louis, Mo., April, 1885, says: Electrical Medication is worth many times the price asked for it. (Send for it.")

J. S. Zerbe, editor *American Inventor*, March, 1884, says in conclusion of a review on Electrical Medication: "Dr. Tipton is to be congratulated for the succesful manner in which he has presented the subject."

———

*D. B. Goldsmith, M. D.*, Ramsey, Fayette Co., Ill., June, 1883: "I have read all the works of prominent electricians of Europe and the United States, but saw nothing based upon common sense and backed up by logical reasoning until I read your book."

*Homœopathic News*, St. Louis, April, 1882 : "Physicians desiring to adopt the battery in their practice will find this a comprehensive and practical work."

---

*American Medical Digest*, New York, April, 1882 : "From reading this work we are impressed with the author's true discipleship and his specific adaptation of electrical medicine to almost every malady to which man is heir."

---

*Minnesota Medical Mirror*, Cambridge City, April, 1882 : "The very liberal and reasonable deductions of the author are convincing at once that he is not riding a hobby," etc.

*American Medical Journal*, St. Louis, Mo., April, 1882 : " The subject of Electrical Medication is presented in a different manner from that to be found anywhere else. In this book we have a practice of medicine characterized by variety—electricity and all therapeutic agents employed by all branches of the profession."

---

*E. O. Neil, M. D.*, Member Royal College Physicians and Surgeons, England—See review in the work.

---

From the *G. E. Medical Journal*, Atlanta, Georgia, 1882, conclusion of review: " Dr. Tipton has presented these subjects of Electrical Therapeutics in a manner that will not fail to interest every reader; nor has he made a humorous hobby out of it."

We could furnish any number of endorsements similar to the foregoing and following:

*Dr. J. A. Dougherty*, United States Medical Examiner for Pensions, says: "The work is worth its weight in gold; entirely unlike anything of the kind that I have ever read."

———

Upon receipt of $3.00, I will send the book to any address by return mail, post-paid. After sixty days, if you do not consider the work worth ten times the price asked for it, return the same and get your money, less transportation.

A. W. TIPTON, M. D.,

TOPEKA, KANSAS.

———

FOR SALE BY BOOK DEALERS.

# PHILOSOPHER AND INFIDEL.

Sir Isaac Newton, the celebrated astronomer of the seventeenth century, was greatly interested in the statements of the Prophet Daniel, and declared his belief that in the fulfillment of them, human knowledge would so increase that men would possibly travel at the rate of fifty miles an hour. Voltaire, the noted French infidel, got hold of this statement and scornfully remarked: " Now look at the mighty mind of Newton, the great philosopher who discovered the law of gravitation ; when he became an old man and got into his dotage he began to study the book called the Bible, and in order to credit its fabulous nonsense he would have us believe that knowledge of mankind will yet be so in-

creased that we shall by-and-by be able to travel
fifty miles an hour! "Poor dotard." (Millennial Dawn, volume 3, pages 63-64. Published in
Allegheny, Penn., by the Tower Publishing Co.)
There has been progression beyond Newton's
expectations. In '92 steam cars run 80 miles
an hour. In '93, at the exposition in Chicago,
miniature electric cars will run 100 miles an
hour. Later on passenger cars will run 150
miles an hour.

Those desirous of further information, see
the arguments for and against in Ludwig:
"Lehrbuch der Physiol." Todd and Bowmans's:
"Physiological Anatomy." Longet: "Traite de
Physiologie." Funke: "Lehrbuch der Physiol."

This fact concerning the nervous fluid has
been one of much dispute among eminent
physiologists, both in this country and in Europe.
We do not consider that this is absolutely beyond dispute, yet, after examining most all the

works extant on the subject, and experimenting much with Electricity, both on dead and living subjects, we are fully convinced of the truth of the above statement. Electricity is the motive power, Carpenter's Human Physiology to the contrary notwithstanding. The following occurs on page 269 of the above named work, referring to the heart and circulation: " It is not possible to imagine that it has any other relation than this to their function; since the formation of each separate element of the organ, of which that of the entire organ is the aggregate, is due to its own inherent vital powers—the supply of blood being only required as furnishing the material on which these are to be exercised. (Nonsense.)

The azotized articles of food yield up their nitrogen for the formation of blood and flesh. But the nitrogen taken into the circulation through the act of respiration is again thrown off. The only or main object of respiration is to produce animal heat and to carry away, as it were, the ashes. Respiration makes room for growth, but in no way produces it; all the material used in building up the system comes to the blood through the digestive and assimilative apparatuses. By the digestive force is meant electricity.

# OLD THEORIES ARE PASSING AWAY.

Science has at last discovered that the sun is not a dead center, with planets wheeling about it, but itself stationary. It is now ascertained that the sun also is in motion, carrying with it its splendid retinue of comets, planets, its satellites and theirs, around some other and vastly mightier center. Astronomers are not yet fully agreed as to what or where that center is. Some, however, believe that they have found the direction of it to be the Pleiades, particularly Alcyone, the central one of the renowned Pleiadic stars. To the distinguished German astronomer, Prof. J. H. Maeidler, belongs the honor of having made this discovery. Alcyone, then, as far as science has been able to perceive, would seem to

95

be "the midnight throne" in which the whole system of gravitation has its seat, and from which the Almighty governs his universe.— *Millennial Dawn, volume III, page 321.*

www.ingramcontent.com/pod-product-compliance
Lightning Source LLC
Chambersburg PA
CBHW021409090426
42742CB00009B/1070